Game Theory

A Beginner's Guide to Game Theory Mathematics, Strategy & Decision-Making

John Cummings

Table of Contents

Introduction

Two million years ago, in a cave, a collection of prehistoric human families came up with a bright idea.

Everyone needed a mix of two things: dependable staples such as carbohydrates and minerals, and rarer still, important nutrients such as proteins. However, the only source of proteins available to them at the time was from dangerous fast-moving animals.

How were they to get them? Each individual knew they could collect enough staples throughout the day to ensure their personal survival even with workflow interruptions from personal domestic duties, animal attacks, and similar hindrances. However, due to the lack of protein, they would never feel at their best. This was even more true for the elderly, who couldn't collect as many staples for themselves and thus wasted away faster as they aged.

Yet, if anyone wanted a fully balanced diet, that'd mean taking a day off to go hunt animals on their own, which usually wouldn't lead to much success.

On an individual level, hunting for protein would almost always mean losing out on a day's worth of staples with nothing in return. Because of this high risk and steep cost, very few people did it despite the necessity of protein to survive healthily.

It soon became clear that everyone providing only for themselves and their children wasn't the best way forward.

So, one day, the elders from all existing families got together and began creating a plan. The plan they eventually agreed upon was that the most physically fit and athletic members of each family would devote themselves to hunting animals every day. Among everyone left, the eldest would devote themselves to looking after the children and tending the hearth throughout the day, while the younger but less athletic would devote themselves solely to foraging and the collection of staples from just after sunrise to near sunset.

Because a small group of people could now look after all the children for most of the day [much like a proto-daycare], everyone else hanging around near the homestead could now spend more time foraging without distraction, leading to much bigger payoffs that allowed them to provide consistently for the whole community. These payoffs would only increase as the extra time devoted to foraging also meant the foragers would become more efficient at finding and collecting food.

This led to a "safety net" of staples, which brought a sense of peace to the athletic group devoted to hunting animals. They now knew that they wouldn't starve if they failed to catch an animal on any given day, and this allowed them to become calmer, quieter, and more patient. All these are the qualities needed in a successful hunter. They would become even more successful as the lack of stress and need to rush allowed them to carefully observe potential prey, and they'd have all the time of the day to learn more and more about the food they wanted to catch. In time, their community would find itself with more protein than they ever could've dreamed of, and all the hunters made things safer for everyone too, allowing the first two groups to work more safely and efficiently with even fewer distractions.

The elderly could eat food that they didn't catch in exchange for teaching children. The foragers could eat protein they didn't hunt in exchange for providing staples. The hunters could rely on staples they didn't forage, so long as they devoted themselves to hunting protein and protecting against animal attacks. This is one of the earliest examples of cooperative game theory in action, even if our prehistoric cavemen ancestors didn't realize it at the time.

Although game theory as a salient concept is relatively young, the principles behind game theory have for millennia allowed humans to anticipate the behavior of their fellows and work out compelling exchanges that suit everyone's self-interest, allowing people to benefit both individually and collectively. In our caveman example, the elderly gained a more secure position and better overall life expectancy, the children would learn more due to the greater experience being passed on. In turn, this would lead to more educated hunters and gatherers. The hunters and gatherers could now devote themselves entirely to what they were best and most efficient at, rather than wasting energy on other tasks that they weren't as well-suited for.

As history moved along, game theory principles would go on to inform strategists, entrepreneurs, politicians, analysts, and philosophers, and help them make strong decisions in military, commercial, social, and moral areas. In all things, game theory principles can be used to logically minimize risk, reduce the chances of calamity, and gain an understanding of rational human behavior.

Whether you're trying to help your team cooperate or you're ambitiously trying to figure out your next campaign strategy, game theory can be a useful guide for you in all arenas of the modern world.

Chapter 1: What is Game Theory?

After that vivid introduction, you likely have a fair idea of what game theory is used for. Further examples include enabling meaningful cooperation, economizing and saving energy without reducing output, and much more.

But what *is* game theory, exactly?

Game Theory Fundamentals

Definitions & Explanations

Game Theory

Game theory is a mathematical model that is used to predict exchanges between rational individuals. This can include cooperative exchanges, such as in our caveman example, but it also includes predictions for how competing groups are most likely to behave toward one another.

For instance, game theory might predict that hunters from another tribe will attack hunters from our first tribe if they're both hunting the same herd of animals. The likelihood of an attack would increase as the herd itself becomes smaller or harder to share, but would decrease according to cultural values (e.g. how sacred human life is in the eyes of the second tribe) as well as the threat of reciprocation (e.g. whether the first tribe or even another tribe might punish or retaliate in response to an attack, and whether or not the second tribe is aware of this possibility).

Factors such as fear or trust can also influence the likelihood of an attack; the greater the mutual trust, the less likely it is that a tribe will choose to make a pre-emptive or first strike.

Game theory works best when payouts are easily known, and consequences can be quantified even if the quantification is just a simple binary "yes or no" or "will or will not."

Rational

Game theory runs on the assumption that all players are *rational.* In the context of game theory, rationality is defined as three things:

1. The awareness of all possible events or outcomes according to one's information set in a given game (e.g. when making a move in chess, being aware of all moves your opponent could possibly make as you can see all their pieces and thus their potential moves are in your information set)

2. The ability to recognize the likelihood of each outcome

3. The desire to maximize *expected* utility based on the likelihood of outcomes (players will want to ensure the highest chance possible for a favorable outcome and minimize the chances of unfavorable outcomes)

Note that being "rational" is not the same thing as being "wise" or "intelligent." It is fully possible for a rational being to have mistaken beliefs regarding likelihood, like believing a plane is more likely to crash than it actually is, or, believing that one is less likely to be mugged or get sick than one is. A rational being might also erroneously recognize the game being played, especially if they're not well educated on game theory.

In Game of Thrones, *the honorable yet straightforward Ned Stark plays a very different game to the snakelike yet cunning Petyr Baelish. Ned believes himself to be playing a cooperation game with Petyr while Petyr is behaving competitively. Both are rational, but Stark's upbringing encourages him to assign an extremely low likelihood to Petyr deceiving or betraying him, while Petyr's upbringing makes him assign an extremely low likelihood to Stark genuinely helping him — which is exactly why Petyr chooses to play competitively instead!*

A rational being isn't predisposed to a specific code, creed, religion, preference, or ideology. A rational being simply makes the best decisions they can according to what they know and believe.

Game

A "game" in game theory is any scenario where the results depend on actions taken by *two or more players*. These can include but aren't limited to, activities that we'd view as "games" outside of a game theory context. In game theory, "games" can include very serious real-world situations where concepts such as winning, losing, and teams aren't always so immutable.

Player

A player is anyone who makes an outcome-determining decision in the context of a game.

Strategy

A strategy involves all the moves a player will make to achieve the best outcome from a game.

Information Set

This is what each player knows at a given point in the game. Information sets can affect the likelihood players will assign to an outcome, and thus, impact their strategy.

Payoff

This is what a player will receive from arriving at a specific outcome. The payoff can be good or bad and can come in the form of currency, utility, or satisfaction. Most players consider personal harm (whether physical or financial) or even death itself to be of negative utility and satisfaction, and thus, do all they can to minimize the risks of such an event in any given game.

More generally, players will consider fulfilling their personal values or ideals to be a positive payoff and going against them to be a negative payoff. In game theory, all players are assumed to ultimately be self-interested, but a self-interested player who highly values their friends or family will seek outcomes that favor them *as an expression of that self-interest.* They simply get a bigger personal payoff from helping their loved ones than they do from quantifiably benefiting themselves directly. Being self-interested is not automatically the same as being "selfish," but in all cases, people will prioritize what benefits either them or their passions and loved ones over anything else.

Equilibrium

When all players have made their decisions in a given game and a specific outcome has been reached, that game is now said to be in *equilibrium.*

The Nash Equilibrium

The Nash Equilibrium is achieved when ALL players have no desire to change their strategy, even when each one is aware of what the other players will do. In these cases, everyone has technically achieved what they wanted with each player getting the ideal payoff according to their values and beliefs. The Prisoner's Dilemma [which will be covered later] is a great way of showcasing the Nash Equilibrium. The Nash Equilibrium is NOT a given in any particular game, but some more complex games might have several resolutions that meet the expectations of the Nash Equilibrium.

An example of Nash Equilibrium being achieved would be in our caveman example from the introduction. Each faction or party got the payoff they wanted and stuck to their strategy, even after knowing everyone else's. The elderly got better diets, protection, and food security in exchange for dealing with children most of the time, hunters got better food security and more time to perfect their craft in exchange for actively pursuing said craft, and gatherers got better diets and protection in exchange for spending more time foraging.

If everyone can get what they want, while only sacrificing things they're willing or able to give, a Nash Equilibrium is likely to be achieved.

Game Theory in Ancient History

Game theory, as we know it today, first began to be expressed as its own distinct concept in 1944 by economist Oskar Morgenstern and polymath John von Neuman. By 1950, it received one of its first game-changing contributions from a mathematician named John Nash (from whom the Nash Equilibrium gets its name). By the 1970s, game theory's expression was well-developed enough to become the pivotal and greatly respected tool that it is today.

However, as our caveman example has hopefully illustrated, game theory principles have nonetheless been around for much longer.

Philosophers such as Plato and Socrates applied a sort of proto-game theory to explain military principles of their time.

Imagine, for a moment, an ancient Greek hoplite on the front lines waiting with his comrades to rebuke an enemy assault. As a player, it may occur to this hoplite that if his comrades are likely to succeed in rebuking the enemy, then the likelihood of his individual presence being necessary is actually quite low. Yet, if he stays, he knows the likelihood of dying would be much higher than if he left. In other words, he gets a bigger payoff from deserting than he does from staying. The defense succeeds either way, but in this case, leaving will minimize his chance of getting hurt.

It may also occur to this hoplite that if the enemy is likely to succeed in their attack, then the likelihood of getting injured or killed by staying increases greatly, again for no point since the battle is lost whether he is there or not. Now, the hoplite has rationalized that no matter who wins or loses, he's better off if he deserts.

However, the problem is that all the other hoplites are players too, and as rational beings all in the same situation, they've also undergone this exact thought process. Now, they realize that if multiple people begin to leave, then chances of defeat and death rise higher for those remaining. Since leaving has the biggest payoff, this convinces all the hoplites that their comrades will abandon them just before the battle, causing a panic, and suddenly the entire hoplite division begins to rout in mass desertion before the enemies show themselves.

Obviously, the commander of the hoplites is a player too, and this outcome frankly terrifies him. To prevent this situation from occurring in the future, the commander makes decisions that encourage his men to stay. For instance, he might choose to make desertion a crime punishable by death, meaning leaving now carries as much risk as staying. Or, he might choose to reward those who perform valorously in the defense. This means that staying now has a potential payoff.

In real life, ancient Greek city-states often made a hoplite's citizenship depend on their service, which greatly increased incentives for staying.

In later centuries before cannons became widespread, this also explains why some commanders would choose to burn their boats upon arriving into enemy territory. By making desertion into safety an impossibility, even the most cowardly soldiers realize their comrades aren't going anywhere and will be able to march into battle more confidently.

If the enemy sees these boats being destroyed, even better. After all, no one would burn their own boats unless they were sure they would win. Witnessing this, then, causes enemy soldiers to overestimate the likelihood of their own defeat. This, in turn, increases their likelihood of desertion which, in turn, increases the likelihood that their defeat truly will come about.

Game theory principles aren't just a matter of acting according to one's rational assessment of potential profit and loss, but also adapting one's actions according to what one can reasonably expect others to do. Those who can count on their comrades will be brave even if they're a coward individually. Furthermore, even the brave will retreat if they think they're just going to get abandoned and die pointlessly.

Outside of a military context, game theory becomes an excellent way for any sort of leader to get a team of diverse rational beings to work together. When punishments and rewards are clear, they can change the way rational players will approach a given game. In team management, the trick is to get each individual to intimately link their personal gain with the collective gain of the group.

Types of Game Theory

Cooperative Game Theory

One of the more common forms of game theory, cooperative game theory involves coalitions of players instead of individuals. In this type of game theory, we see how different types of groups will interact with one another when only the payoffs for each outcome are known. In addition to assuming all players to be rational and self-interested, cooperative game theory also assumes that a coalition will do what it can to enforce cooperation within itself. The challenge of the coalition in cooperative game theory is to fairly divide the payoff of an outcome among all of its member-players. In the caveman example, the foragers need to decide how much starch and minerals to keep for themselves and how much to share in exchange for childcare, protection, and meat, while all other groups in the caveman coalition likewise need to make similar decisions regarding how they split the "payoffs" gained through their cooperative strategy.

Non-Cooperative Game Theory

Non-cooperative game theory is the most common form of game theory and is one of the simplest. It is game theory at its base form, simply dealing with how rational agents will interact with one another to achieve their own personal goals. One of the simplest forms of a non-cooperative game would be hinging a bet on Rock-Paper-Scissors. Two individuals each with three options compete to win the payoff.

A less simple example would be two rivals in a company vying for a specific promotion. Before being able to act as a group, an aspiring leader will need to consider the non-cooperative games individuals may be playing and find ways to encourage the behaviors needed for cooperation.

Game Theory in Parenting: Enforcing Cooperation

Say a parent, as a reward for something wonderful their children did, places chocolate on a table, and two siblings with equal love for chocolate see it.

Assume that, when it comes to their children, the parent always keeps their word. If they promise a punishment or reward, they always follow through with their children. This means their children trust them, and that their words are "game-changers" that define the rules of any game they're involved in.

Now, the parent says if the chocolate is shared, the children will be rewarded with another one in the future. The first child to ask nicely will get the chocolate, and thus gets to decide how to break it up and share it.

In this situation, although there is an incentive to share, there is no incentive to share more than the bare minimum, so it becomes a competition to see who can ask the nicest first. If a child believes they can get away with surreptitiously riling their sibling up, and thus force them into a tantrum that bars them from asking nicely for chocolate, then they also have the incentive to do this, since it is to their advantage *in this particular game* and they might not have the foresight to be playing anything long term at the moment.

However, say the parent adds another rule; although the first child gets to break up the chocolate, it is the *second* child who decides which piece the first child actually gets, and refusing to honor this decision will bar the first child from having any share of the next chocolate.

Now, the siblings are encouraged to cooperate. The first child now knows that if they split the chocolate unfairly, they'll simply get left with the smaller piece (or with none at all in the future if they're obstinate), and so they do their best to split the chocolate as evenly as possible with their sibling. This disincentivizes underhanded tactics and encourages the two individual players to now see themselves as a coalition.

From this point, the question becomes less about "Who gets the chocolate?" and has a better chance to evolve into "What can we do to create a scenario in which we can share chocolate?" lifting the game from non-cooperative into cooperative territory.

This game can get more complex of course. For example, a child might deliberately choose a strategy that leads to a smaller piece because they get more utility out of their sibling's gratitude than they do out of sweets, but for the sake of simplicity, we won't be diving into all that just yet.

Symmetric/ Asymmetric

A symmetric game is one in which all players (or coalitions if the game is cooperative) have access to the same options, are going for the same payoffs, and the anticipated payoffs from a particular strategy do not change just because the players change.

Most sports without defined team roles can fall under symmetric game theory assuming there are two teams of similar skill. Aside from that, most symmetric games are actually very short-term or emphasize tactics over strategy. For instance, chess might seem to be symmetric at first but quickly becomes asymmetric as lost pieces directly impact the options available to the player.

An asymmetric game can be one where options are not the same for both sides, but can also be one where objectives aren't the same (e.g. hide & seek has two sides with different payoffs, and as such behave differently). It can also be one where objectives and options are the same, but players can take on non-negligible specializations or different strengths and weaknesses that nonetheless influence how they select options or pursue objectives (e.g. civilizations in Age of Empires).

Our cavemen are a great example of a game that turned asymmetric through specialization.

Simultaneous/ Sequential

A simultaneous game is one in which a player must make a decision without knowing what all other players will do. This is generally because the decisions are being made at the exact same time. Again, look at Rock, Paper Scissors, which is a great example of a simultaneous game. Another great example would be the Prisoner's Dilemma which we're covering later in this book. In simultaneous games, players still know what they stand to win or lose in any given outcome and are still aware of what outcomes are possible. The rules of the game are still known. Anonymous voting, such as in most public democratic elections, is an example of a simultaneous game.

By contrast, sequential games are those where each player takes turns and chooses their actions in response to one another in order to reach the desired outcome. A player might not know what their opponent will do next, but they still know what said opponent *can* do. Chess is a great example of a sequential game. In politics, voting without anonymity (e.g. in Parliaments or Congress) can be turned into a sequential game.

For example, three politicians are called to vote on whether or not to approve a pay raise bill that'll benefit them but increase the burden on taxpayers. Two votes out of three are needed to pass the bill, and all three benefit if it passes. If an individual votes for the bill, they will face public backlash, but if the vote fails, all three lose potential utility. In this hypothetical scenario, the players believe that the raise would grant more utility to themselves individually than the public support would (e.g. the raise is large enough to fund a campaign that could recoup double the support lost), although the utility of the public support itself is still significant.

A cunning player might wait for his colleagues to vote yes before voting no so he can reap the benefits without the backlash and thus get the biggest payoff. Of course, if all are equally cunning, one might choose to vote "no" first, thus forcing the other two to take the public backlash in order to secure the raise for greater net utility. From here, it becomes a situation of who can ensure that their turn is first.

In a similar but more complicated scenario, the player who desires public support most might have set up a system of promises and favors to the other two beforehand to further incentivize their colleagues to vote "yes," but by that point, our players are no longer playing simply one game.

Sequential games give players more room to change according to the choices of other players.

Finite/ Infinite

This is the most interesting, but one of the most complex components of game theory. A finite game is simple. It has a certain number of players, each with a certain number of options according to a fixed set of rules, and the game itself cannot run beyond a certain limit. Almost all sport matches and parlor games, whether symmetric or asymmetric, are finite.

An infinite game would be something like the Cold War where new players can come in whenever they wish, where options and rules have room to shift and evolve, and where it never truly ends. Some players simply get exhausted, take their ball and go home, but new ones come in whenever they wish (Sinek, 2016).

Commerce and economics, when taken as a whole, are another example of an infinite game. Individual businesses are also able to play finite games simultaneously according to their needs. There is no inherent limit on how many games a player may participate in at once, whether finite or infinite.

The possibility of playing an underlying infinite game or being motivated by an ulterior finite game is why a nation may choose to pass anti-corruption laws or why an organization might check you for conflicts of interest before letting you participate in a trusted role. In order to play a cooperative game (e.g. working with a business or government towards a specific outcome), other players in a coalition need to know they can trust you, and they will find it harder to do so if you have unfettered incentives to work against them in any of the myriad games you may be playing.

When law or preliminary checks weaken the more collectively harmful ulterior incentives you may potentially have, you become easier to trust, even if the other players do not know you that well personally.

Chapter 2: Where Can Game Theory Be Used?

By now, you likely have an excellent idea of how widely game theory can be applied. This chapter will discuss the most common areas that game theory is used in.

Resource Allocation

Game theory principles can aid greatly in resource allocation, either through encouraging resources to be allocated for the collective good of a coalition, or through gaining insights into how a given player may split resources when left to their own devices.

We can examine how resource allocation behaviors emerge through a variety of games.

The Dictator Game

In this game, two players are present. One is assigned the role of "dictator" and the other is not. The dictator is given a large sum of utility (e.g. in the form of money or another reward) and is allowed to split it with the second player as they see fit. The second player has no real ability to affect the first player's decision. According to an article on behavioral economics, about 50% of dictators will keep all the utility to themselves, 5% of dictators will split it evenly, and 45% will give the second player a smaller share assuming this is a once-off interaction between the two players who never encounter one another again.

Note that in our earlier example with chocolate and siblings, if the parent didn't put checks in place around how the chocolate should be shared, the children would have essentially been playing a Dictator Game.

The Ultimatum Game

This is very similar to the Dictator Game but comes with a caveat. It is fully possible for the second player to *reject* whatever split the first player proposes. In this game, if the split is rejected then *neither* player gets anything. This guarantees that the first player will at least offer *something* to the second. In the real world, personal relationships play a huge factor in this game. If relations are strained or if the two players are normally in competition, player two may decide it's worth not getting anything if it means player one doesn't get anything either, especially if they know that player one is going to get the most benefit out of the deal should it be accepted.

In this case, the onus is on player one to ensure that the utility they're offering to player two exceeds the effective utility player two would get from punishing player one through denial.

How Adding Additional Players Affects the Ultimatum Game

In the real world, a variant of the Ultimatum Game can occur where multiple players are making offers to player two. In this scenario, player two can only choose one of the offers (e.g. player two has to decide who to co-found a business with).

Assume all players are aware of one another. Each player competing for player two's acceptance knows that player two will accept the offer that is most in their favor. This incentivizes all the other players to create splits that heavily favor player two. If three or more players are simultaneously competing for an accepted proposal from player two, then they're all likely to offer exceedingly favorable proposals.

Making an Ultimatum Game recurring – that is, the players can keep moving on to new partners over and over until they reach a proposal they like, can have a similar effect. However, in this case, the player who is favored isn't always player two. If lots of people are seeking to make offers with few people around to accept them, then people in player two's shoes will be favored. However, if there are lots of people waiting around for offers, yet not lots of people proposing them, then it is those who are in player one's shoes who have an advantage.

If this sounds familiar to you at all, it may be because it's beginning to sound a lot like basic supply and demand. The more difficult someone believes it is to get a payoff at all, the more likely it is that they'll settle for a less favorable payoff than they otherwise would (e.g. paying ruinous amounts for bread when reasonably priced loaves become unobtainable, or selling a house for below your ideal price due to a lack of offers).

How Reputation and Recurrence Affects the Ultimatum and Dictator Games

Even in games where only one player is splitting resources with another, the presence of other people in the form of confidantes can still affect the game. Studies have found that when player one is aware that player two can comment on player one's generosity or criticize them for their greed, player one becomes more likely to split resources in a cooperative manner. Now, player one knows they can gain significant utility in the form of widespread trust or appreciation (Wu et al, 2019). This becomes even more true in games where the players expect to meet each other again at a future date, where potential utility can also exist in the form of reciprocation. If this sounds familiar, this may be because rationales like this have led to classic schoolyard adages like "Treat others the way you want to be treated."

Interestingly, this same study found that while reputation does encourage cooperation, the threat of rejection or punishment only ensures it for as long as the threat remains. Unlike positive reinforcement, negative reinforcement becomes ineffectual the moment a player recognizes it can no longer be enforced. In some cases, negative reinforcement can even reduce cooperation. If a player's reputation is lowered too much, they may feel alienated from the group entirely and thus see little incentive in trying to integrate anymore. This phenomenon can be used to rationally explain the benefits of forgiveness.

Sociology

Resource allocation is deeply entwined in many sociopolitical issues. It should therefore be no surprise that just as game theory can apply to resources, so can it apply to sociology itself.

The Volunteer's Dilemma

The volunteer's dilemma is a game of two or more players where an unpleasant task needs to be done. Everyone benefits if the task is done, and the worst possible outcome occurs when no one steps up.

A common example would be performing a household chore. If you do it, you devote time and energy to getting yourself dirty in exchange for a more pleasant and hygienic living space.

A more poignant example would be reporting crime. Say you are on the street when you witness an assault. In this assault, multiple gang members are brutally attacking a single unarmed innocent. Aside from you, there are fifteen or so other witnesses. Trying to stop the assault or even conspicuously running off to call for help could put you in considerable danger. Yet, if everyone witnessing this event does nothing, the whole neighborhood becomes more dangerous as future assaulters will now know that there's no risk attached to the payoff of their crime.

All the other witnesses realize this too. As rational beings generally want to minimize risk and maximize payoffs, no one wants to be the one who runs off to call for help. So, they wait and see if someone else does instead. If this sounds familiar, this may be because it links to "Bystander Syndrome," where people in a random crowd become less responsive to emergencies than they would be if they were alone. The more people there are in the crowd, the longer the delay will be as everyone will be waiting for *someone* to do something without knowing who that *someone* will be.

A more white-collar example would be reporting fraud within your company. In this scenario, imagine you know that nearly everyone in the accounting division is engaged in fraud, yet somehow upper management does not know. You would report it, but your colleagues know too. You don't really want to be the one who gets most of the accounting division fired. You don't want to deal with the potential problems being a whistle-blower can cause down the line. So, you hope that one of your colleagues reports it instead.

The problem is, your colleagues have likely all had the exact same thought process as you. You become more aware that if the fraud keeps going unreported, then it's highly likely the company will fold, costing *everyone's* jobs. Everyone is in the habit of waiting for someone else to do something.

Coming full circle back to resource allocation, in a household with limited internet connectivity, someone needs to volunteer to abstain from data usage so the rest of the house can benefit from it.

To overcome the inertia caused by Bystander Syndrome in a Volunteer's Dilemma, there are several things we can do. A community can make a schedule beforehand, deciding who is "on shift" to volunteer when an emergency comes up. In our assault example, this is the reasoning behind having a neighborhood watch. Once an individual is explicitly (albeit temporarily) made responsible for reporting crime, they can then be armed with backup and protective gear to minimize their personal risk and maximize their likelihood of reporting without delay.

This is also the reason why many countries have witness protection programs and why whistle-blowers are often protected with anonymity. Minimizing the risks of being a volunteer in a Volunteer's Dilemma increases the likelihood of someone stepping up, as the common good becomes less fettered as a motivating force. Note that training an individual to be more competent in a specific emergency is a valid way of minimizing risk.

However, increasing incentives outright (e.g. paying whistle-blowers for making a claim) can be a double-edged sword as it also incentivizes deliberately causing issues to later "solve."

That said, for something domestic such as taking out the trash, something as simple as showing appreciation can go a long way to safely incentivizing a player to volunteer. Gratitude is a powerful utility in relationships that are already based on love or friendship and can be surprisingly enticing as a payoff in such situations.

Animal Behaviors

This is an interesting application of game theory as not many of us can imagine animals as rational. However, game theory can nonetheless be used to understand why symbiotic relationships form between seemingly unlikely species.

For instance, the fragile and the diminutive cleaner wrasse fish might choose to enter a shark's mouth. The shark, in turn, does not eat the fish even though it easily can. Why is this?

For the sake of this example, consider the fish and the shark to both be players, rational beings with a desire to maximize utility. The shark *could* eat the wrasse, but the wrasse has very little meat on it. Despite the meal being a very easy one, the utility gained from eating it is still relatively small, but this alone doesn't prevent it from being eaten.

What prevents the wrasse from being eaten is the fact that it provides an alternative utility. It cleans parasites, dead skin cells, and rotting leftover meat out of the shark's mouth. The utility of this is especially valuable because while the shark could gain ample food utility from other fish species, it can only get cleaning utility from this one. Therefore, for the sake of maximizing utility (in this case, through having access to multiple forms of meaningful utility), a shark will not eat the wrasse as doing so will deny it a utility that it cannot get elsewhere.

In this way, the shark gets the best payoff from letting the wrasse live. However, live skin and mucus are better for the wrasse than dead skin and leftovers (Meiden, 2015). So, why doesn't the rational wrasse choose to maximize its payoff and take a bite out of its client?

The moment the wrasse hurts the shark, the shark assigns a negative utility to keeping the wrasse alive. After all, any rational being will do all it can to minimize harm to itself, and, leaving something alive to eat you from the inside is no good. At this point, the only positive utility the shark can gain from the wrasse would be through eating it before more damage can be done. Which the shark then promptly does.

The rational wrasse, realizing this, decides it cannot enjoy its ideal payoff if it's dead, and thus, assigns a negative payoff to any notions of eating live shark flesh or mucus. This leaves eating the dead flesh and parasites as its best realistically possible payoff. Live shark flesh simply isn't good enough to justify the risk of death compared to safely eating parasites.

The shark and wrasse now reach a Nash Equilibrium, where the shark adopts a strategy of not eating the wrasse, and the wrasse adopts a strategy of only eating the unwanted things in the shark's mouth. The two species expect this behavior from one another, and it leads to the best possible outcome for both, a unique utility for the shark and a free meal for the wrasse.

Through this, a symbiotic relationship is born.

The principles behind this example of animal behavior can be extrapolated to humans too. Offering utility in a form that is unique yet helpful can open doors for you that might have otherwise remained closed. This is why novelty and innovation are important for entrepreneurs, especially those trying to break into an otherwise saturated market.

Is That All?

No, of course not! A lot of these applications extend into other areas. Resource allocation extends into business and politics which we'll cover later in this book, while sociology can branch into morality and philosophy. Being able to predict the most likely choice someone will take in a situation, as well as understand *why* that choice was taken, is absolutely essential for any philosopher who wishes to do good and bring prosperity through their proposed moral structures and codes. Understanding what people are likely to do and what they're motivated by can give vital inroads into creating practices that both appeal to them *and* better them.

Game theory is also especially helpful for creating AI programs, as understanding game theory can help such programs make decisions that provide utility to their users.

Finally, going back to morality for a moment, game theory is great for puzzling out the best outcomes in scenarios where we're experiencing both internal and external conflict. In game theory, this is known as a Mixed Motive Game.

In a Mixed Motive game, there are compelling incentives for two players to both compete against and cooperate with one another. To resolve the game, then, requires the player to come to terms both with their own motives as well as with what they believe the other player will do.

One of the best-known Mixed Motive games, and indeed one of the most famous scenarios in game theory overall, is what is called the Prisoner's Dilemma, which is pivotal enough to demand its own chapter.

Chapter 3: The Prisoner's Dilemma & Encouraging Cooperation

The Prisoner's Dilemma is an amazing example of game theory because it teaches us several things:

- What is good for the individual isn't necessarily good for the collective.
- What is good for a selfish individual doesn't necessarily remain good once other players begin behaving equally as selfishly.
- The ability to communicate is pivotal for ensuring cooperation.
- Splitting a group up before putting its individual members under pressure is a manipulation tactic that is highly useful for subverting said group.
- Trustworthiness is an incredibly valuable utility in scenarios where communication has been disrupted.

The premise of the Prisoner's Dilemma is simple. Two players, which we will call Todd and Viola, are captured by an authority figure. This authority can be part of a regime that is beneficial or oppressive, but what is certain is that the authority is *trustworthy*. Both Todd and Viola believe their captors will keep their word and will not lie.

Now, although Todd and Viola are in custody due to suspicion, the authorities are not able to pin an actual crime on them. Todd is asked to confess Viola's role in a crime. If Todd does this, he will be allowed to walk free without further repercussions. Viola, however, would be sentenced to five years in prison.

If Todd chooses not to confess, then he'll be detained for one year on trumped-up charges before being let go.

Todd is then told that Viola has been offered the exact same deal.

If neither Todd nor Viola accuse each other of crime, then both will be detained for one year on trumped-up charges before being released. If they *both* accuse one another, they'll be imprisoned for two years each, the rationale being that after a mutual accusation, they're now both "undeniably" guilty, but have gotten a lighter sentence due to their "honesty."

So, at first, it seems that the optimal strategy would be for both to refuse confession.

However, that is sadly not the case. The system has specifically been set up to encourage confession, which you'll see now as we break things down.

If a player does not confess, they face either one year of prison or five.
If a player *does* confess, however, they face either *no* imprisonment or *two years* of imprisonment.

In other words, whether Viola confesses or not, Todd realizes he'd be better off confessing as well. Viola, being a rational being, also realizes this and realizes she'd be safer overall if she confessed too.

This scenario echoes with the hoplite story we went over earlier. The moment there is any doubt as to the trustworthiness of one's comrades, the tendency of the player is to default to looking after themselves first and foremost, even at the cost of letting down the group overall.

Both players want to minimize their prison sentences as much as possible. They could both achieve this if they simply kept silent. However, this isn't the same situation as the one with the shark and the wrasse. In that aquatic scenario, the outcome with the biggest payoff also carried by far the highest risk of harm. In our prison cell, however, the outcome with the biggest payoff, getting to walk free, is also tied to the *lowest* risk of harm. Two years *instead* of five would be the worst possible outcome.

So, already, confession is highly tempting. It is so tempting that both parties confessing is actually the Nash Equilibrium for this game if we do not presume on any morality outside of rationality.

In the real world, that temptation might disappear if Todd and Viola had a chance to speak privately, but their captors won't give them that chance. By preventing the prisoners from knowing what each other's plan is, there is room to plant a seed of doubt.

That seed grows and begins to eat at them. So, gnawed by the terrible feeling that they're going to get betrayed, both Todd and Viola rat each other out to rule out the possibility of getting stuck with a five-year sentence.

While it is true that this strategy completely prevents either of them from getting the worst possible payoff in the Prisoner's Dilemma, it also means they're now both stuck with the *second-worst* instead, whereas they could've both had the *second-best* instead if they worked together.

Relevancy of the Prisoner's Dilemma

The Prisoner's Dilemma is not only relevant for fugitives or the victims of an oppressive regime. Healthy societies are based on people trusting one another and being somewhat open to personal risk for the sake of the common good, as the Volunteer's Dilemma from earlier may have implied.

For this reason, understanding the Prisoner's Dilemma becomes relevant for any situation where the best collective outcome demands personal risk where the selfish choice is the most profitable individually, and where everyone being selfish leads to a worse outcome both collectively and individually than all the players being cooperative with one another.

Encouraging Cooperation Among Players in the Prisoner's Dilemma

Now, it should be noted that even though playing selfishly in the Prisoner's Dilemma is the dominant strategy due to the perceived safety net it provides, not every player will have an innate desire to play this way. Why? While all players are rational, as human beings, our players are often *social* as well. They perceive utility in aiding a common good, or in trying to achieve the best collective outcome with their fellow players.

Amazingly, in a test where over forty players played once-off Prisoner's Dilemma games with one another anonymously, it was found that there was a 22% chance of cooperation overall (This Place, 2014). This is despite the lack of communication, prior history, or attachment between players. Even more amazingly, it was found that while the number of people who always betrayed was *significantly* larger than those who always cooperated, as well as much larger than those who cooperated 50% of the time, the constant betrayers were less than half the size of those who tried cooperation *at least once*. This tells us that the majority of people (about one in four, give or take) start out as idealistic and will give strangers a chance at least once before growing cynical from betrayal.

However, you do not know how much betrayal a person has gone through nor do you know how likely they are to let go of a difficult past. That said, you do not want a difficult past to become a roadblock in cooperation. So, how do we encourage people to work with us?

Iterative Prisoner's Dilemma

In a normal Prisoner's Dilemma, which is only played once, the safest strategy certainly seems to be a betrayal. On the other hand, what if the game repeats?

For instance, consider two students who are assigned to work together on a school project. "Betrayal" here means choosing to neglect the project, while "cooperation" means choosing to put time, effort, and research into it. The results of each combination are similar to our specific example: by betraying, you either receive a good grade for no effort (maximum payoff), or a poor grade for no effort (second-worst payoff; assume the student uses the time to do a better job on other assignments or is able to do revision work later, but the loss of marks still impacts them).

In contrast, cooperation means either a good grade for fair effort (second-best payoff) or a good grade for huge effort (worst payoff;

assume the necessary effort of doing double the intended work would mean concurrent tasks from other subjects are completed to a poorer standard, or the student develops detrimental health or stress issues over this time period).

However, it is rare that a school only assigns one group project over the course of a year. Normally, there are several. So, if a player chose the "betrayal" strategy for the first project, what do you think will happen in the next project?

The betrayer, as a rational being, now realizes that everyone else in the class is more likely to betray him in turn so that they aren't taken advantage of.

Players will outright refuse to work with him when they can or will otherwise behave in a way that annihilates any certainty of the betrayer gaining their ideal maximum payoff. In other words, the betrayer has now locked themselves into a situation where they can only ever get the second-worst outcome as opposed to the second-best they could've gotten had they cooperated from the beginning.

Luckily, most rational players have some measure of foresight and anticipate that this will be their fate if they ever choose to betray. Therefore, when a player needs to work with the same group of people over and over again, they are more likely to make decisions that favor the common good.

Communication & Trust

Another key component in cooperatively playing the Prisoner's Dilemma is communication and trust. If the players are allowed to speak to one another, they are able to clarify their intentions and reassure one another that they have each other's backs. However, this alone isn't always enough. For instance, imagine that two competing companies are deciding whether or not to spend more money on advertising. They realize that if they both increase their marketing budgets, then all that'll happen is they'll expend massive utility in terms of time and money for no net gain, as the competing advert campaigns will simply cancel each other out.

To avoid this, they decide to speak to one another and try to reach an agreement to keep their marketing budgets as is to avoid unnecessary escalation. However, even if an agreement is made, what stops one of the businesses from now increasing marketing anyway, stealing all the customers away from the business that kept its word? This is an especially poignant thought if one of the businesses has itself been suckered in the past. Now we're back to square one, and betrayal is once again the safest strategy.

This shows that communication is meaningless without trust. Trust, in turn, is built by being known for keeping your word, honoring your deals, and respecting your agreements. If both players have a reputation for holding their word as their bond, at least when it comes to previous dealings with one another, then cooperation becomes immensely easy the moment both parties swear to their deal.

In the case of Todd and Viola, if they had previously both taken personal risks to aid one another multiple times in the past, then they could both resist the temptation to betray simply by promising to look after one another's interests even when they're separated.

Building trust so that employees will look out for the collective good even when they aren't being watched is the rationale behind team-building activities. Other ways of determining trust include letting a potential player borrow something of yours that has high utility to them, yet low utility to you, and seeing if the object is returned by the agreed date. Setting up multiple low-risk Prisoner's Dilemma scenarios can help build the trust needed to confidently tackle trickier ones e.g. grouping employees together on a scavenger hunt gives them a chance to express cooperation and build confidence in one another before handling a more serious assignment.

Trust becomes incredibly important in Iterative Prisoner's Dilemmas, as it can provide utility in terms of additional contacts who wouldn't have risked working with you were it not for your reputation.

Tit for Tat

When two players or coalitions can trust one another's word, the "Tit for Tat" strategy can be effectively employed to further encourage cooperation (Hayes, 2020). In any iterative game, a player can promise "Tit for Tat." This promise means two things. Firstly, "if you scratch my back, I'll scratch yours," and secondly, "an eye for an eye; if you provoke me, I'll retaliate in ***equal*** measure."

We already saw shades of this strategy in both the school project and marketing campaign scenarios where it was shown to be both a powerful deterring and reassuring force. "Tit for Tat" can be attempted even before trust is fully established if the promising or threatening player seems reasonably capable of carrying out their actions, but it is most powerful when the player being coerced has no doubt that the promises will be fulfilled, one way or the other. Failing to fulfill on a "Tit for Tat" promise can damage trust, although there is a caveat. If you're already well trusted, then lightening a punishment from what was promised can be seen as mercy which can sometimes garner appreciation. Reducing a reward from what was promised, however, can only be seen as stingy, and will always damage the trust.

Note that trust does not inherently mean either a friendly or a tense relationship. It just means one where both parties can reasonably expect one another to do what they claim they will do. In more positive relationships, the threat part of "Tit for Tat" may be made...

- subtler, like threatening to refuse a favor unless it is reciprocated (...well, I'd love to lift the tariffs off your imports, Mr. President, but until you do the same for my country, I'm afraid I cannot help you).

- more playful (...I swear, if you get drunk and vomit on my wedding dress, you'll be footing the laundry bill).

- or, if appropriate, even omitted entirely.

When employing punishment or retaliation according to tit for tat, do not forget that just because it looks reasonable within the context of the game doesn't mean it'll always look fair to outside observers. Take care not to create ripples of resentment outside of what was intended, as is often the case when punishment is harsh beyond precedent. As an example, think of how family feuds are unwittingly started when one goes too far settling a personal matter.

Institutional Incentives

Organizations can also provide top-down incentives to ensure cooperation in a Prisoner's Dilemma. In a scenario where the players are literally prisoners, perhaps the prisoners are part of a cartel. In this case, the cartel might warn of harsh punishments for anyone who chooses to confess ("Snitches get stitches"). Of course, the arresting police department might then counter with a witness protection program to level out the playing field again.

In our school project example, an education system might be structured so that peers in a project evaluate one another's contributions and must give brief reports of each other's performance. This adds extra incentive to cooperate, as now players who opt to be lazy cannot get the same amazing mark as the player who opted to stock up on caffeine and put in the hard work. This naturally leads to a more balanced division of labor where there's less incentive to force any member into overworking themselves. Some educators may even encourage students to report unhelpful partners, bringing in the reputation mechanic early and making cooperation a more urgently needed strategy.

For our business scenario, where they were discussing marketing plans, it's possible they may choose to make their agreement legally enforceable or else elect to both join a guild that will enforce the terms of the agreement, ensuring cooperation and fettering the incentives for betrayal.

Note that institutional incentives do not work unless the relevant players within the institution feel like they can trust it. This is why some institutions can seem so pigheaded about enforcing rules. Maintaining trust in their system is more important to the institution than being liked or being seen as cool or relaxed. Without trust, the institution cannot continue as an effective coalition, effectively leading to its death, an outcome that it'll try as hard to avoid as any other player.

Social Norms & Societal Values

Related to institutional incentives are the social norms of a culture. If a player is raised and conditioned throughout their childhood to favor the good of the group, then they're more likely to place themselves in personal risk to achieve the best outcome for their group. This is because fulfilling personal beliefs counts as a valid form of utility, while going against a personal belief counts as a negative utility. This further reduces the net incentive of betraying the group, while increasing the incentive to take risks for it.

Note, however, that players conditioned this way are not machines. They are still rational and will do what they can to minimize self-harm. Remember, the more often a player has been betrayed, the more likely they are to default to betraying in turn for the sake of their own safety, at least until trust has been established.

A societal value that can bolster cooperation further would be that of valuing the future. Even if a player does not feel they can trust strangers and is placing heavy emphasis on putting their own safety first because of this, they might still choose to cooperate because they know that the more they betray, the fewer people they can eventually establish meaningful trust with thanks to prior relations or poor reputation. This will encourage the player to take the risk of cooperation even after prior betrayals, although when and how often this risk will be taken depends on how well they believe they can weather getting betrayed again.

In real life, a player may opt to simply play games involving betrayal as little as possible, or they might choose to move somewhere else and try to start fresh.

One of the biggest mistakes we tend to make as players is that we fail to properly observe the norms of other players, and make assumptions based on appearance, nationality, sexuality, or ethnicity. This can lead to undeserved levels of trust when overestimating someone's similarity of norms to yours, while leading to undeserved distrust or phobia when underestimating someone's similarity of norms to yours. While norms are a helpful tool, presuming on them can unfavorably skew the way you play any given game.

Chapter 4: The Shapley Value

Although there are many steps one can take to encourage cooperation. At the end of the day, they will struggle to hold your coalition together unless every contributing player feels as if they're being reasonably compensated for the effort that they're putting in.

The Shapley Value, Explained

The Shapley Value is a concept invented to try to resolve reward disputes. The idea is that, by fairly distributing the loads *and* the gains of a venture among all members of the coalition, they'll be more inclined to work with each other.

Note that "fair" does not necessarily mean "equal" unless all players have access to the exact same time, resources, initial utility, and motivating forces. As you can imagine, such a team does not occur in real life, and the Shapley Value is used explicitly to work out reasonable rewards in situations where contributions to a project were unequal.

In our caveman example, the Shapley Value would've been a great help in determining just how gains should be split.

What is a reasonable or fair reward? In the context of game theory, the Shapley Value should help provide everyone in the coalition with a reward of more significance than what each player would have gained from working alone.

This is because players have no rational reason to collaborate if they do not gain from it. For example, if you could sell $30 worth of hot dogs a day when running a stand by yourself, but would still only make a net total of $30 each day when hiring an assistant, why would you keep that assistant? Why would you spend time and money trying to manage and worry about another person, expending all that extra energy and effort for no extra gain?

The Shapley Value, Applied

If you were to Google the Shapley Value, you might see that its equation form is a headache to express and understand, even among some mathematicians. Regardless, we can explain things easily enough through an example, and then through covering the step-by-step process in simplified terms (Knight, 2014).

Imagine you and two friends are going out to a nightclub. In this hypothetical situation, you'd normally all get there by hiring your own taxis. You live a moderate distance away, so your fare would be $15. One of your friends (let's call her Janine) lives much further away so she'd have to pay $45. Your other friend (let's call him Sebastian) lives very close and would only need to pay $10.

However, you have the bright idea of realizing that if you all took a single taxi, the fare would just be $45 split between the three of you. Collectively, this will save you $25 (or earn you $25, whichever way you want to look at it).

What's the best way to split the bill so that everyone feels like they've saved money without freeloading or owing favors?

You could try to divide $45 by three and that'd give you a very neat $15 per person. Janine would love this, but Sebastian wouldn't, and frankly, you don't really have an incentive to keep riding together like this either, especially since you do all your socializing once arriving at the nightclub itself. So, when dividing "equally," only one person out of the three is actually happy with the arrangement.

This may be fine at first, but unless either you or Sebastian are deeply in love with Janine, this arrangement doesn't provide enough utility to everyone to be sustainable and will likely cease if the relationship as a coalition or the players as individuals receives any difficulties.

How, then, do we split by Shapley Value? First, we must determine the marginal contribution of each player, as well as each subgroup. Individual marginal contribution can be calculated as:

Individual Marginal Contribution = Subgroup Output - Subgroup Output Without Individual

The Individuals

We already know that, alone, you would pay $15, Janine would pay $45, and Sebastian would pay $10. We can express this mathematically as:

v(c) =

- 10, if c = (S)
- 15, if c = (YOU)
- 45, if c = (J)

Here, *c* stands for *coalition.*

Subgroups

We also know that, due to the way this taxi operates, the amount each subgroup pays would equal what the highest individual fare would be. This isn't true for all scenarios. These initial numbers should be gathered by observation, not calculation. Mathematically, this is expressed as:

v(c) =

- 15, if c = (S, YOU)
- 45, if c = (S, J)
- 45, if c = (YOU, J)

The Grand Coalition

Of course, all three of you together would still only pay $45 collectively, which can be expressed as:

v(c) =

- 45, if c = (S, YOU, J)

Altogether, that's:

v(c)=

- 10, if c = (S)
- 15, if c = (YOU)
- 45, if c = (J)
- 15, if c = (S, YOU)
- 45, if c = (S, J)
- 45, if c = (YOU, J)
- 45, if c = (S, YOU, J)

Now we can start working out the marginal contribution of each individual. We start doing this by seeing how much each individual would pay toward the output of the grand coalition ($45, the amount the three of you would pay collectively) if they had to go first, second or third.

To properly work out marginal contribution and thus determine the Shapley value, the rule is that the first person must pay as much as they can towards this sum without going over what they would normally pay when alone (e.g. if Sebastian goes first, he'd pay the full $10, just as he would when alone).

The second person must then pay as much as they can, but without creating a sum larger than what they'd have if they were just in a subgroup with the first person (e.g. if you pay second, you'd only pay $5, rather than the $15 you'd pay when alone because the normal sum you'd have when in a subgroup with Sebastian would total $15. Since Sebastian has already paid $10, paying more than $5 would go over that and create a non-optimal situation).

The last person to pay must then cover the difference (e.g. Janine, if she pays last, pays $30 to ensure the cost of the grand coalition is covered).

To summarize our bracketed explanations: if Sebastian had to pay first without knowing what everyone else is going to do, he'd happily put forward $10 as his marginal contribution just as he would when alone. Now, you come along. You only pay $5, instead of the full $15 you'd pay when alone, because Sebastian went first and

v(c) =

15, if c = (S, YOU)

Finally, Janine pays and covers the rest of the cost at $30. She's individually saved $15 here, so she doesn't mind. You've saved $10. Sebastian has saved nothing, so we know we haven't arrived at the Shapley Value yet.

The bill split we've just witnessed here can be expressed as:

(S, YOU, J)→(10, 5, 30)

To arrive at the Shapley Value, it is important that we first gather more data by repeating this payment process with every possible permutation.

(S, YOU, J)→(10, 5, 30)

(S, J, YOU)

(YOU, S, J)

(YOU, J, S)

(J, S, YOU)

(J, YOU, S)

Through this, we determine what each person's marginal contribution would be in each variant of the same situation.

Once we repeat the bill split with every permutation, we see:

(S, YOU, J)→(10, 5, 30)

(S, J, YOU)→(10, 35, 0) [Remember, $v(c) = 45$ *if* $c = (S, J)$, so Janine effectively covers the whole bill after Sebastian if she goes second and you get off without paying anything]

(YOU, S, J)→(15, 0, 30)

(YOU, J, S)→(15, 30, 0)

(J, S, YOU)→(45, 0, 0)

(J, YOU, S)→(45, 0, 0)

With this information, we're now ready to calculate the Shapley value for each member.

This is done by finding the average of each person's marginal contribution.

Your marginal contributions are 5, 0, 15, 15, 0 and 0. In case high school math classes are a fuzzy memory, the average of this can be calculated by summing those values together, then dividing them by the number of different values. In this case: (5+0+15+15+0+0)/ 6

= (35)/6

=

5.833333333333333333333333333333333333333333333333333333333333333333

≈ 5.83

This is your Shapley Value for this scenario and means your optimal contribution to the taxi fare during an average ride would be $5.83. You have individually saved $9.17. Your fare has also effectively been cut down to about a third of what it'd be if you were paying alone. Amazing!

Now, for Sebastian...

(10+10+0+0+0+0)/6

= (20)/6

=

3.333333333333333333333333333333333333333333333333333333333333333333333

≈3.83

Sebastian has a Shapley Value of 3.33, and his optimal contribution would be $3.33, saving him $6.67 individually. His fare has also effectively been cut down to a third of what it'd be if he were paying alone. Amazing!

Finally, Janine...

(30, 35, 30, 30, 45, 45)/6

= (215)/6

=
35.8333333333333333333333333333333333333333333333333
33333333333333333

≈ 35.83

Janine has a Shapley Value of 35.83. Although $35.83 is nowhere near a third of the $45 she'd pay if alone, she has effectively saved $9.17 individually, so she is still much better off than if she were riding alone.

Now, to make sure there were no mistakes, let's add up the Shapley Values and see if they equal to the bill that needs to be paid.

35.8333333...+3.3333333...+5.8333333...
= 45

In real life, some decimal values aren't always fun to work with. There is no 33.33333333... cent coin, and, using rounded down Shapley Values would lead to underpaying the taxi driver (by one cent). So, in a scenario like this, your coalition might choose to slightly doctor the results.

If all players agree, there is nothing wrong with deviating a little from the Shapley Value (e.g. ignoring the sub-coalition component to alter the bill into something like $5, $7.50 and $32.50 for Sebastian, you and Janine respectively). The Shapley Value isn't gospel, but it is a great base to use for determining optimum contributions.

When it comes to splitting bills by default, the Shapley Value system favors those who are not budgeted to make huge contributions while still benefiting those ready and capable of performing the heavy lifting. It creates the air of generosity while still profiting the primary "giver" and thus reducing chances of resentment from either side.

When it comes to *generating value*, however, it ensures that those who make strong, consistent contributions to a group venture reap the highest rewards without disincentivizing those who still need to catch up, incentivizing everyone to put forward as much as possible.

Chapter 5: Game Theory in Business & Politics

Many of the principles already covered in this book are highly applicable to both business and politics, but we will now take a look at a few more specific examples here.

Business

The Coordination Game

People like being trusted. Although treating every game like a potential Prisoner's Dilemma by shoring upon incentives, readying promises, and so forth can be seen as wise, it can take a lot of time and energy to always be on guard. Sometimes you need to know when you can relax, and the coordination game helps us spot exactly when.

A coordination game is any scenario where the best possible payoffs are achieved from following the same course of action.

Imagine that two extremely influential leading health technology companies are now deciding between introducing an innovative surgical instrument that could potentially earn them hundreds of millions of dollars in profit, or, choosing to improve an older machine that would earn them far less.

Neither company has the resources to make this new instrument a household name among hospitals on its own. So, if either company tried to introduce this technology independently, the product's obscurity would mean very little profit would be generated at first.

In this scenario, if the companies both chose to introduce this new instrument and worked together to promote it in the same year, then both companies would earn $800 million over the course of that year.

If only one company introduces the technology that year, however, then it'll only earn $200 million in profits for that year. The company that didn't introduce it in that year, however, will have trouble catching up and staying relevant in later years, and thus risks making $0 net profit in its expensive attempts to catch up later.

If neither company introduces the new technology, then they'll both be able to make comfortable $400 million profits off their revised older machines instead without danger from going obsolete.

Here, if one chooses to introduce the technology, the best-case scenario is a profit of $800 million, while the worst case is a profit of $200 million.

If one chooses not to introduce, the best case is now only $400 million profit at best or $0 at worst.

So, both companies want to release this technology, but it's important that they release it in the same year, both to avoid being left behind as well as to maximize profits.

Unlike the Prisoner's Dilemma, the Nash Equilibrium here favors cooperation. Because of this, you know you can afford to be friendlier and focus more on fostering goodwill during your negotiations between your company and the competing one. Unless its leadership is exceptionally hateful or petty, it likely won't desire to quarter its own profits just for the sake of denying yours (see the Ultimatum Game from Chapter 2), and it especially won't risk making zero profit just to dent yours.

When you identify a coordination game, it is easy to get the second player to go along with you simply by reminding them of the huge benefits. If they try to make unfair demands of you in exchange for cooperation, you can always remind them that cooperation works out the best for both of you, making it easier for you to reject unfavorable ultimatums.

Price Cutting

Game theory in business allows you to anticipate price-cutting from competitors through the lens of the Prisoner's Dilemma (see Chapter 3). This is an interesting case of the Prisoner's Dilemma because, while both companies refusing to undercut each other ("cooperation") leads to a more favorable collective outcome for our players than engaging in a price-war ("betrayal") would, the collective of outside observers watching the game, in this case the consumers, would feel differently. Consumers love it when both companies opt for "betrayal" in this variant of the Prisoner's Dilemma because it means they now get access to much cheaper goods, which in itself can benefit society on a greater collective level.

This is a reminder that games affect more than just the people playing them and that what benefits the playing collective might not benefit the external non-playing collective. Although the two companies are now saddled in their second-worst outcome through "betraying" (undercutting) one another, the consumer base has now achieved their *best* possible outcome from this game (great deals) and they weren't even playing!

Competition between two players can create surprising benefits and opportunities for third parties. When taking the external collective into account, the Prisoner's Dilemma ceases to be a symmetric finite game and becomes asymmetrical and infinite. This is why economics is said to be an infinite game overall; it's never just about the companies, but also about the millions of people they serve.

Politics

We looked at a possible application for game theory in politics when we touched on sequential games and how they can affect voting in Chapter 1, but now we are going to look at something a little more interesting.

The Deadlock Game

Have you ever wondered why Congress doesn't seem to be able to make up its mind? When we hear of a "deadlock" we tend to think immediately of an indecisive vote, but the Deadlock Game's most famous application has nothing to do with voting at all. It's to do with nukes. Rather, it's to do with why disarming them is so difficult for world leaders, even when literally everyone else would be on board with it.

This scenario is like the Prisoner's Dilemma, but...

- Both parties get the *second-worst* outcome if they both cooperate.
- Both parties get the *second-best* outcome if they both defect.
- A party gets the *absolute worst* outcome if it cooperates but the other defects.
- A party gets the *absolute best* outcome if it defects when the other cooperates.

The reason why this is called a Deadlock Game is that both players benefit the most on an individual and collective level if they refuse to cooperate. In a nuclear disarmament context, the reasoning is that both countries want to keep stockpiling nukes in order to remain on par with one another, while being able to lord over the rest of the world (second-best outcome). If they both cooperate and disarm, then they still remain on par with one another, but risk diminishing their role as a global nuclear superpower (second-worst outcome). If only one cooperates and disarms while the other defects and chooses to keep stockpiling, then the one who cooperated will find themselves unable to threaten retaliation against their still-overwhelmingly-armed opponent (absolute worst outcome for them) while the one who kept stockpiling now has a total military advantage over their opponent and can now dictate terms more easily (absolute best outcome for them).

In this scenario, how do we incentivize cooperation? Both sides might demand huge asymmetric concessions from one another. However, another reliable way is through *changing the game*. We do this by changing the values of the players themselves.

In the context of disarmament, there are two key values to look at; strategic and economic. The less a player depends on overwhelming military force in order to cement themselves as a superpower, the more likely they are to disarm. The less a player depends on being a superpower at all in order to accomplish its wider goal, the more likely it is to disarm. The more expensive maintaining their nukes become, the more likely they are to disarm.

After many years of playing the Deadlock Game, assuming they no longer depend on overt or overwhelming military action to support themselves, both players might decide that it simply isn't worth spending millions to maintain weapons they're never going to use. At that point, stockpiling becomes a negative payoff, not a positive payoff, and the Deadlock shifts into a Prisoner's Dilemma.

From here, it is now just a matter of establishing trust, and disarmament agreements can now realistically proceed. Historically, this trust was established between the USA and the USSR through repeated conferences or negotiation "games" where each side would keep its word.

A more relevant Deadlock scenario for the modern-day would be climate change. Imagine two competing countries trying to reach an agreement on cutting down on emissions. Doing this would negatively impact their economic output as decreased environmental exploitation would lead to lower/slower profits. If only one country keeps to the agreement while the other doesn't, then the cooperator will become more vulnerable to economic invasion from the defector.

As the two countries aren't necessarily superpowers in this case, however, cooperation can be encouraged through an external enforcer to protect cooperating countries from defection. This is why, in real life, the United Nations tends to preside over international agreements to cut down on greenhouse gases.

Another factor that invariably changes the game is when the exploited resources begin to run out. The scarcer a resource becomes, the less payoff there is in continuing to mine it exploitatively.
At this point, reducing pollution may carry a bigger payoff than running emission-heavy industries at full steam, but by then there is also likely to be little environment left to save, especially if the perceived value in reducing pollution is low, as that'll mandate resources to be even scarcer before conservation or emission reduction seems like an attractive option, potentially leading to a "Tragedy of the Commons" where a shared renewable resource (e.g. fish, drinkable water, breathable air) is virtually destroyed through overharvest or exploitation.

This shows the importance of trustworthy international organizations in incentivizing conservation.

Foreign Policy

Imagine that two nations have a disagreement. Nation A is doing something that Nation B doesn't like. It could be an unfair tariff on international trade, it could be the annexation of a neighbor, or it could even be research into WMDs. Nation B then threatens Nation A, telling Nation A that Nation B will act unless Nation A's current course of action ceases. This action could involve sanctions, revoking trade deals, or even causing war. To keep it simple, let's say that Nation A has the choice to "Continue" or "Cease" while Nation B has the choice to "Intervene" or "Back Down".

To see what each nation will really do, it's important to first know what their objectives are and what they're capable of doing. For this example, assume Nation A's primary goal is to become an international power while Nation B wishes to prevent this.

Suppose Nation A believes that even if Nation B intervenes, Nation A's current action cannot be entirely thwarted (e.g. if building WMDs, being attacked won't lead to all its WMD facilities being destroyed).

So far, it seems to be in Nation A's best interest to then continue, no matter what Nation B does. However, its secondary objective is to avoid intervention on the part of Nation B regardless, as being sanctioned or attacked is still bad news.

In this case, Nation A's best outcome is if they can continue without being attacked, it's second-best being if it continues despite being attacked, it's second-worst being that it stops and thus avoids being attacked, and its absolute worst outcome being if it stops and gets interfered with anyways (or if Nation B turns out to be stronger than expected and forces them to stop through intervention directly).

Nation B, however, cares less about intervention than it does about Nation A stopping, but, is still more than willing to do so if it encourages Nation A to cease. Therefore, Nation B's best outcome is if Nation A ceases without Nation B needing to intervene, its second-best is if Nation A stops after Nation B intervenes, its second-worst is if Nation A continues and Nation B backs down, and its absolute worst is if Nation B intervenes and Nation A manages to continue in spite of this (as now Nation B has agitated A, who has grown in power anyways, while B has weakened itself spending resources on intervention).

So, to sum up:

- If Nation A ceases, it can only experience its worst outcomes.

- Assuming Nation A's assessment of Nation B's capabilities are accurate, A can only achieve its best outcomes from continuing.

- If Nation B intervenes, it experiences either its second-best or absolute worst outcomes.

- If Nation B backs down, it experiences either its absolute best or second-worst outcome.

So, in this scenario, Nation A realizes it gains a better outcome from continuing no matter what B does, and B realizes it gains a better outcome from not intervening no matter what A does. Therefore, the threat is ineffectual and, as a side-effect for future games, Nation B has also lost some trust.

You can play around with priorities to see under what conditions these two players will view stopping or intervening as their best decisions.

Some general things to consider:

- The more likely Nation B's intervention is to forcibly stop Nation A's action in its entirety, the more inclined Nation A will be to stop without intervention (so as to avoid its absolute worst outcome). This incidentally means that the more likely Nation B's intervention is to achieve total victory (at least in the perception of Nation A), the more likely it is that Nation B will achieve their best outcome (getting what they want without needing to intervene first).

- The more Nation B cares about being trustworthy, the less likely they are to back down.

- External nations may join the game, perhaps funding (and thus incentivizing) intervention on the part of Nation B, or they might give sanctions of their own to increase pressure to stop on Nation A.

- Threatening B to back down or funding A to continue would normally be wasteful expenses for an external nation to consider, as that's what A and B would've chosen anyways without outside intervention.

- However, funding A could help it weather through sanctions from other nations, while threatening B could help scare it into not intervening even if other nations would reward it for doing so.

This is all heavily oversimplified, but it does show how game theory can be used to model interactions between nations as well as what might tip them towards one decision or another.

Chapter 6: Case Studies

The MAD Doctrine: Cold War

The MAD doctrine is an interesting component of the Cold War as it ensured that the USA and USSR seldom, if ever, fought against one another directly. It also ensured that major world powers would never again enter into large-scale open warfare (with each other, that is) after the close of 1945.

The USA and USSR, both being rational players, care about preserving themselves and will do all they can to stay alive.

However, at one point they realized that they both have *ridiculously huge* arsenals of city-destroying weapons. They also realized that they both have automated systems to ensure retaliation no matter what. So, whoever uses their nuke on their opponent first receives equal if not greater devastation. That retaliation would in turn trigger another automated system to retaliate against *that* strike, meaning both players automatically wipe themselves out if either one of them launches a nuke against the other.

Neither side is able to disable the other's nuclear weapons entirely (a.k.a achieve a "Splendid First Strike"). This is because the method of retaliation is via a *secure* second-strike such as intercontinental ballistic missiles, plane missiles, *and* submarine-based missiles hidden in the ocean used in tandem.

This point was reached because both sides kept playing the Deadlock Game, and as such, naturally chose to keep stockpiling nukes as their dominant strategy.

Eventually, both players realized that if they started a serious armed conflict with one another, the costs would be significantly worse than if they simply accepted an unfavorable peace with one another instead.

This is a more complex form of the foreign policy method we described earlier, but it essentially is a situation where both sides are guaranteed to reach their absolute worst outcome if they fight. For this reason, they are now significantly more likely to make peace with each other in areas they wouldn't consider before, as an unfavorable peace is still preferable to a nuclear holocaust, and neither player will consider maximizing their outcomes in a way that risks this absolute worst outcome.

To illustrate the point, no one would risk defecting in a Prisoner's Dilemma if getting nuked was the worst possible outcome for doing so, even if the best outcome for doing so would mean becoming a world dictator.

Why Racial Tensions & Segregation Keep Making Comebacks

A video game called "Parable of the Polygons" has been used to intuitively highlight Tom Schelling's game theory model of neighborhood segregation.

If you want to test it out for yourself, you can find it here: https://ncase.me/polygons/

The premise of Schelling's model is that even mild preferences for similarity or homogeneity can lead to deep divides in society over time.

In this video game, half the population is triangles and the other half are squares. Funnily enough, none of them are actively or hatefully racist, so already they're off to a better start than many of us. In fact, they *prefer* living in diversity! Yet, invariably, deep segregation arises. How?

Each polygon is an individual player with three payoffs:

- Living in a diverse neighborhood brings maximum payoff (an optimal mix of familiarity and discovery).

- Living in a neighborhood where everyone is like you brings a neutral payoff (you get on with people, some are pretty nice, but it can feel kind of dull or static at times).

- Living in a neighborhood where less than one-third of your neighbors are like you gives a negative payoff (you don't like feeling surrounded or being in a local minority).

So, if a polygon is a part of an especially small minority in its neighborhood (e.g. comprises 33% of the neighborhood or less), then it will move in order to get a better psychological payoff (this system, due to assuming a diversity-loving society, does not account for forced removals).

So, starting with a random jumble of shapes, some start happy, while others want nothing more than to move. So, they all start moving in reaction to one another, and, as the game shows, they end up settling into situations where some live in happy diversity, but most live in homogeneity.

Why is this an issue? Recall Chapter 3 and the importance of trust in cooperation. If you never interact with a group (or at least an assortment of individuals that you perceive as a group), then you never have an opportunity to build trust. Without trust, it is more likely that one or both parties will behave selfishly when in a space of potential competition with one another. This then perpetuates a vicious cycle of pain and distrust.

So, in this situation, it seems that settling for a neutral payoff isn't enough. Even in a world that loves diversity, even something as small as a 33% bias for one's perceived kind can lead to a social divorce between groups. Reducing this bias to even 10% does not change the situation much.

Now, you might be tempted to say "Well, the neutral shapes should then be moving too". This would be pretty close to the mark. However, in real life feeling "meh" on its own isn't always enough motivation to pick up sticks and go through the hassle of moving to another area.

What if we change the rules in another way?

- Living in a diverse neighborhood brings maximum payoff (assume payoff is largely psychological or even value-based in this scenario).

- Living in a neighborhood where more than three-quarters of your neighbors are like you gives a negative payoff.

- Living in a neighborhood where less than one-third of your neighbors are like you gives a negative payoff

If you apply a rule like that in your own "Parable of the Shapes" game, you might see something profound. Avoiding segregation and promoting tolerance isn't just a matter of loving diversity (Bliss, 2014). It's also a matter of being adventurous, of being curious, and of abhorring echo-chambers of any kind, even the ones that have diversity as their chief interest. This is what lays the foundation for seemingly different groups to work together in the absence of MAD.

Conclusion

As we reach these last few parting words, one should be reminded that as a player, you can only choose what action *you* can take. At no point in game theory [or in real life, for that matter] can you perfectly control or determine what another person absolutely will do. The best you can hope for is to give them reasons to do one thing or another.

To play any game, however, requires more than just an objective. It requires knowledge. When you know what those around you can do, when you know what they value, and when you know what they strive for, then you can know how best to work with them or, if necessary, how to protect yourself against them.

The games shown to you in this book, although relatively simple, have given you a taste of how useful game theory can be for understanding and encouraging behaviors among the people we differ from.

It is easy when studying game theory to grow a little depressed or to become a little more cynical or Machiavellian which this particular book has attempted to balance by hugely emphasizing game theory's use in cooperation.

Should you ever read further, diving into colder texts, never forget that the Prisoner's Dilemma is not just an imaginary game in an imaginary jail cell. It is a game all of us play every day. Although cynicism could indeed protect oneself from the worst of all possible outcomes, the Dilemma shows it can also lock us into the *second-worst*, while the study mentioned along with it shows us something profound: the more often we betray, the more likely others will give up on being social and solely play as *rational*, becoming betrayers in turn.

Use game theory to create the life you want to lead. Understand what your own payoffs are, and whether or not your biases work with or against them.

Then, try to understand people's needs and use appropriate incentives to push things in the direction you wish. The more people can be aligned with one another through mutually beneficial payoffs, the more they'll work together, and the closer we'll get to experiencing the collectively-best of all possible worlds.

As you've now learned about the Shapley value and the subjectivity of payoffs, I'd like to end this section off with a little bit of wisdom derived from economists such as Adam Smith and Milton Friedman (Goodreads, 2019).

That wisdom is this: one of humanity's greatest achievements came about when it realized that, through a willing exchange, both parties are now able to profit rather than one at the expense of the other. The global resource pie is not fixed, nor has it ever been.

If even cavemen could get that right, can't we?

References

Anderson, K. (2017, June 9). *Teaching Game Theory and the Tragedy of the Commons in Middle School.* Population Education. https://populationeducation.org/teaching-game-theory-and-tragedy-commons-middle-school/#:~:text=Game%20Theory%20In%20Ecology%20With%20Shared%20Natural%20Resources&text=This%20phenomenon%2C%20called%20The%20Tragedy

Bliss, L. (2014, December 10). *An Immersive Game Shows How Easily Segregation Arises—and How We Might Fix It.* Bloomberg. https://www.bloomberg.com/news/articles/2014-12-10/an-immersive-game-shows-how-easily-segregation-arises-and-how-we-might-fix-it

Chen, J., Lu, S.-I., & Vekhter, D. (1999). *Other Dilemmas.* Cs.Stanford.Edu; Stanford University. https://cs.stanford.edu/people/eroberts/courses/soco/projects/1998-99/game-theory/dilemma.html

Dagklis, K. (2020). *Game Theory: Two Real World Examples.* Studying Economics; The Economics Network. http://www.studyingeconomics.ac.uk/blog/game-theory-real-world-examples/

Fisk, P. (2019, January 2). *The Infinite Game ... Most games have finite boundaries, and measures of success ... but not business.* GeniusWorks. https://www.thegeniusworks.com/2019/01/the-infinite-game-most-games-have-finite-boundaries-and-measures-of-success-but-not-business/

Goodreads. (2019). *Milton Friedman Quotes* (Author of Capitalism and Freedom). Goodreads.Com. https://www.goodreads.com/author/quotes/5001.Milton_Friedman

Guner, S. (2012, June 21). *A Short Note on the Use of Game Theory in Analyses of International Relations*. E-International Relations. https://www.e-ir.info/2012/06/21/a-short-note-on-the-use-of-game-theory-in-analyses-of-international-relations/

Hayes, A. (2020, June 1). *Game Theory* (B. Barnier, Ed.). Investopedia. https://www.investopedia.com/terms/g/gametheory.asp

Kenton, W. (2019, November 13). *Shapley Value*. Investopedia. https://www.investopedia.com/terms/s/shapley-value.asp

Knight, V. (2014). Cooperative Games and the Shapley value [YouTube Video]. In *YouTube*. https://www.youtube.com/watch?v=w9O0fkfMkx0

McAdams, D. (2017, December 18). *Game Theory and Cooperation: How Putting Others First Can Help Everyone*. Frontiers For Young Minds. https://kids.frontiersin.org/article/10.3389/frym.2017.00066

McNulty, D. (2019, November 13). *The Basics Of Game Theory*. Investopedia. https://www.investopedia.com/articles/financial-theory/08/game-theory-basics.asp

Meiden, L. (2015, February 20). *Exploitation and Cooperation by Cleaner Wrasse – Shark Research & Conservation*

Program (SRC). Shark Research; University of Miami. https://sharkresearch.rsmas.miami.edu/exploitation-and-cooperation-by-cleaner-wrasse/

Newkirk II, V. (2016, April 21). *Is Climate Change a Prisoner's Dilemma or a Stag Hunt?* The Atlantic. https://www.theatlantic.com/notes/2016/04/climate-change-game-theory-models/479340/

Nitisha. (2015, January 9). *5 Types of Games in Game Theory (With Diagram)*. Economics Discussion. https://www.economicsdiscussion.net/game-theory/5-types-of-games-in-game-theory-with-diagram/3827#:~:text=Even%20in%20case%20of%20interchanging

One Minute Economics. (2016). Game Theory Explained in One Minute [YouTube Video]. In *YouTube*. https://www.youtube.com/watch?reload=9&v=YueJukoFBMU

Picardo, E. (2019, May 19). *How Game Theory Strategy Improves Decision Making*. Investopedia. https://www.investopedia.com/articles/investing/111113/advanced-game-theory-strategies-decisionmaking.asp

Ross, D. (2019, March 8). *Game Theory (Stanford Encyclopedia of Philosophy)*. Stanford.Edu. https://plato.stanford.edu/entries/game-theory/

Saxena, S. (2019, November 11). *Game Theory In Artificial Intelligence | Nash Equilibrium*. Analytics Vidhya. https://www.analyticsvidhya.com/blog/2019/11/game-theory-ai/

Sinek, S. (2016). What game theory teaches us about war [YouTube Video]. In *YouTube*. https://www.youtube.com/watch?v=0bFs6ZiynSU

Spaniel, W. (2013, February 25). *Mutually Assured Destruction (MAD) – Game Theory 101*. Gametheory101.Com. http://gametheory101.com/courses/international-relations-101/mutually-assured-destruction-mad/

Talton, B. (2011, March 31). *Game Balance: Symmetry vs. Asymmetry*. Boardgamegeek.Com. https://boardgamegeek.com/thread/636692/game-balance-symmetry-vs-asymmetry

This Place. (2014). The Prisoner's Dilemma [YouTube Video]. In *YouTube*. https://www.youtube.com/watch?v=t9Lo2fgxWHw

Vitelli, R. (2016, April 20). *Exploring the Volunteer's Dilemma*. Psychology Today. https://www.psychologytoday.com/za/blog/media-spotlight/201604/exploring-the-volunteers-dilemma

Wu, J., Balliet, D., Kou, Y., & Van Lange, P. (2019, March 28). *Gossip in the Dictator and Ultimatum Games: Its Immediate and Downstream Consequences for Cooperation. Frontiers in Psychology*. https://www.frontiersin.org/articles/10.3389/fpsyg.2019.00651/full

Zyga, L. (2006). *A "prisoner's dilemma" for real-life situations*. Phys.Org. https://phys.org/news/2006-09-prisoner-dilemma-real-life-situations.html

CPSIA information can be obtained
at www.ICGtesting.com
Printed in the USA
LVHW051112130323
741492LV00016B/1264